YOUR KNOWLEDGE HAS VALUE

- We will publish your bachelor's and master's thesis, essays and papers

- Your own eBook and book - sold worldwide in all relevant shops

- Earn money with each sale

Upload your text at www.GRIN.com and publish for free

The Raw Material Concrete. Formation Steps and Applications of Concrete

RC. Nivita

Bibliographic information published by the German National Library:

The German National Library lists this publication in the National Bibliography; detailed bibliographic data are available on the Internet at http://dnb.dnb.de.

ISBN: 9783346826619
This book is also available as an ebook.

© GRIN Publishing GmbH
Nymphenburger Straße 86
80636 München

Print and binding: Books on Demand GmbH, Norderstedt, Germany
Printed on acid-free paper from responsible sources.

The present work has been carefully prepared. Nevertheless, authors and publishers do not incur liability for the correctness of information, notes, links and advice as well as any printing errors.

GRIN web shop: https://www.grin.com/document/1329267

CONCRETING: BASICS AND APPLICATIONS

Abstract: Besides water, cement, sand, stone aggregates, and water are the four main ingredients of a concrete mix. RCC stands for "reinforced cement concrete," which is formed by incorporating a steel rod cage into the concrete mixture. This paper explores the formation steps and applications of cocnrete. While being transported and placed, fresh concrete shall remain stable and not separate or bleed when subjected to the forces associated with limited types of handling activities. Variations in the concrete's quality have been linked to mistakes made during the batching process, it has been found. When deciding on a mixing technique, it's important to keep the following in mind. A significant quantity of concrete can only be produced if the mixer is properly and regularly maintained. The time it takes to transport concrete from the site where it is mixed to the site where it is deposited should be as short as possible. For the sake of preserving the material's workability and preventing ingredient loss, this is done before placement.

1.   Introduction

Cement, sand, stone aggregates, and water are the four primary components that go into making concrete [1]. Reinforced cement concrete, often known as RCC for short, is produced when a cage made of steel rods is combined with the concrete mix. This results in the creation of the material [2].When subjected to stresses during handling activities of a limited type, fresh concrete should be stable and should not segregate or bleed throughout the process of transportation and placement [3]. This is particularly important. The mixture needs to be cohesive and mobile enough to be placed in the form around the reinforcement [4]. Additionally, it needs to be able to be cast into the required shape without losing its continuity or homogeneity in accordance with the techniques that are available for placing the concrete at a particular job

1

[5]. The mixture needs to be able to be properly and thoroughly compacted into a dense, compact concrete that has a minimum amount of voids under the facilities that are already in place at the site for compaction [6]. A good blend from the perspective of compatibility should be able to fill 99 percent of the voids that were originally there [7]. The steps involved in preparing concrete are outlined ahead.

The components that go into the making of concrete: 1. Cement 2. Fine aggregate 3. Coarse aggregate 4. Water. The process that is necessary for the manufacture and execution of the object to be concreting can be roughly divided into the following stages [8]:

1. Mixing of the components in batches:

It is essential that all of the components that go into the manufacture of concrete are subjected to precise measurement within the margins of error that are allowed in order to achieve excellent consistency throughout the finished product [9]. Material can be batching by hand or using a batching plant, depending on your preference. The amount of work that will be performed and the quantity of concrete that must be produced should both factor into the decision on which batching system will be used [10]. One bag of cement is considered to have a weight of fifty kilogrammes, despite the fact that the weight of batching cement is always measured by weight. When manually batching, the fine and coarse aggregate are measured by volume, but when mechanically batching, also known as using a batching plant, the fine and coarse aggregate are measured by weight [11].

In the event of batching done manually, the quantities of coarse and fine aggregates are adjusted in such a way as to call for one bag of cement for each batch. In this scenario, wooden farmas are typically utilised for the purpose of determining the volume of aggregates [12]. The utilisation of a batching plant is a practise that is carried out for all significant works that include a significant number of concreting labour and call for the maintenance of controlled conditions during production [13]. Batching equipment can range from a simple wheel barrow scale that feeds a portable mixer to a complex automatic or semi-automatic batching system, depending on the volume of the work being done [14]. It has been discovered that variations in the quality of the concrete can be partially attributed to errors that occurred during the batching process [15]. This procedure can be adequately regulated, and its control is superior than that of other aspects that are responsible for the variance in concrete quality [16]. It has been observed that inadequate batching is responsible for more variation to an inconsequential amount of total variation. As a result, the operation of batching is a highly significant activity [17].

## 2.	Type of batching

There are two different approaches to batching, which are as follows:

1. batching based on volume

2. the use of weight in batching

Batching in terms of volume: - In this method of batching, the quantities of the materials are determined according to their volumes [18]. The gauging of cement is not recommended because it is difficult to obtain accuracy in its measurement. The actual volume of a given weight of cement is dependent on how it is filled into a gauge box and whether or not it is shaken down after being filled [19]. Gauging cement is not recommended because it is difficult to obtain accuracy in its measurement. It is possible for the density of the cement to range anywhere from 1.12g/cucm to 1.6g/cucm, depending on how heavily it is packed into the container after it has been poured in [20]. Because of this, using an entire bag of cement on the site, while it may be more convenient, is unnecessary except in situations when bulk cement is used [21]. If only a portion of a bag of cement is needed, then the amount needed should be measured out using an ordinary spring balance and a bucket that is hanging from the balance [22].

It is possible to measure aggregates based on their volume, and wooden batch boxes known as farmas are employed for this purpose. The dimensions of farmas should be such that they can accurately measure the amount of aggregate that must be combined with a whole bag of cement in order to produce the desired mix [23]. They should not be made so big that they are difficult to manipulate in any way. Each bag of cement that is shipped out from a factory is pre-weighed and measured to have a total net weight of fifty kilogrammes [24]. Cement weighs 1.44 kilogrammes per litre, so the amount of 50 kilogrammes would be 50 divided by 1.44, which is 34.72 kilogrammes; nevertheless, for convenience, it is considered as 35 litres. The farma ought to be of a manageable size, with units of measurement that are multiples of 35 litres. It is recommended to use farmas that are tall and narrow rather than ones that are shallow and wide [25]. The dimensions of a farma in India are 40 by 35 by 25 centimetres. It is recommended that the farma be constructed out of timber that has been prepared and is three centimetres thick. In most cases, the volume of water is measured, assuming that the technology used is accurate. It is difficult to accurately quantify the amount of water that is contained in coarse sand that has been wet [26].

Batching by weight: These days, in order to make high-quality concrete, all of the ingredients are measured in terms of their weight [27]. The mixes themselves have weight batching mechanisms, so that the dial gauge that is put into the loading skip can plainly display the weight of any material that has been placed in it after it has been mixed [22]. There are many different types of weight batching equipment available, encompassing a wide range of capacities to accommodate a variety of different sized punches [25] .

4

3.   The process of mixing concrete

The process of mixing concrete is a crucial step. The mixing method should be carefully selected, and the following considerations should be taken into account [28].

THE CAPACITY OF THE MIXER: - The volume of the concrete after it has been compacted is used to characterise the size of the mixer. In the past, the volume of the unmixed ingredients found in their loose state was also stated in the description of the product. The latter can be up to fifty percent larger than the volume after compacting it [29]. Mixers can range in size from 40 litres, which are used in laboratories, to 13 cum, which are used in industrial settings. In order to manufacture a batch of concrete with specific proportions, it is necessary to know the number of 59-gallon bags of cement that will be required for any mixer of a specific size [30].

MIXING TIME: - On a job site, there is frequently the urge to mix concrete as quickly as humanly possible; hence, it is essential to be aware of the minimum mixing time that is required to generate concrete that is uniform in consistency and composition in order to achieve the desired level of strength[18]. In all honesty, the amount of time necessary to mix concrete differs depending on the size and design of the mixer [31]. To be more precise, the criterion for determining whether or not something has been adequately mixed is not the amount of time spent mixing but rather the number of times the mixer has been rotated. In most cases, mixing properly only requires roughly 20 revolutions of the mixer's paddle [32]. Since the producers of the mixer propose a particular speed of rotation that is optimal, it follows that the total number of rotations and the amount of time spent mixing are interdependent [27]. The start of the mixing time occurs when all of the dry ingredients have been added to the mixer, and the water should not be added until after one fourth of the total mixing time has passed. After this point, the mixing time begins to be counted [33].

5

OUTPUT OF MIXER: - This will be determined by the capacity of the mixer as well as the concrete mix that is utilized [12]. The duration of the mixing period and the amount of time it takes to charge and discharge the drum, commonly known as the time of mixing cycle, both have an effect on the amount of product that can be produced in one hour [17]. Even under typical conditions, when delays occur owing to one cause or another, the result is generally a significantly lower output than what was promised by the manufacturer [34].

4.   The process of feeding the ingredients

It is impossible to provide general guidelines for the process of adding components to the mixer because these guidelines are contingent on the properties of the mix and the mixer [35]. In most cases, a lesser quantity of water should be fed first, followed by all of the solid components, preferably fed uniformly and simultaneously into the mixer [13]. Feeding the mixer should be done in this order. If at all feasible, the majority of the water should be given at the same time as the solids, and the remaining portion of the eater should be given afterwards [36].

CHECKING OF MIXER:- Before beginning operation of the mixer, it is important to double check the following criteria [37].

1. The mixer needs to be positioned on ground that is both stable and level.

2. Maintain the speed of the mixer at twenty revolutions per minute during the mixing process.

3. The water controlling device should be checked if the mixer has built-in tanks that are used to supply the water for the mixing process [18].

4. Inspect the mixer drum and the blades to ensure that they are clean and clear of any clumps or concrete that may have adhered to them.

5. Check out all of the different machinery that can move things.

6. When loading the skip, the ingredients should be placed in the following order: coarse aggregate, then fine aggregate, and then cement [21].

After these components have been loaded into the drum, the materials will be deposited in the opposite sequence, with the coarse aggregate being the final component. It will be beneficial to push any sand or cement that may have become stuck to the mouth of the skip [38].

Maintaining the mixer in such a way that it produces a sufficient amount of concrete requires that it be done in an orderly and consistent manner [8]. If there is a batching plant, the supply pipe needs to be examined to make sure there is no choking. This may be done by pushing a normal ball down the pipe when it is under a lot of pressure [17]. The choking danger will be eliminated after the ball is passed. If there is no ball available, then each pipe needs to be removed, cleaned, and then reinstalled [27]. The maintenance of the mixer requires that the following points be taken into consideration:

1. The mixer should not be left loaded with the concrete mix either before the shift is finished for the day or during lunch break. If it is cleaned, then the mixer's age and quality will be negatively impacted [19].

2. Grease or oil should be applied to the internal face of the mixer's drum, as well as the surface of the blades, etc.

3. The outside of the mixer's housing should be clean as well.

4. If the mixer is powered by gasoline, then both the rope and the belt need to be inspected.

When mixing concrete, the goal is to coat the surface of all of the aggregate particles with cement paste and to integrate all of the constituents of concrete into a single, uniform area [24]. When we mix things together, we take extra precautions to make sure that the components don't lose their original state in the process. The mixing of concrete can be done in one of two distinct ways [39]:

a). Mixing by Hand: - - When the quality of the concrete that is going to be used in a project does not justify the use of machinery for mixing the concrete, this method of mixing concrete is brought back into use [11]. The technology is advantageously utilised in locations where the use of machinery is either not possible due to a lack of availability of such machinery or in circumstances where noise is not acceptable. Concrete cannot be mixed effectively by hand because of this. A watertight platform that is approximately 3.5 metres in length and at least 2 metres broad should be supplied for hand mixing [7]. Because the earth and dirt will also be mixed in with the concrete, mixing concrete by hand should be done on the ground. It is recommended that the mixing be done on a hard and clean surface [2].

b) Mixing by Machine: The device that is utilised for the purpose of mixing concrete is referred to as a concrete mixer. Concrete must be mixed by hand in many nations, which results in higher labour costs [25]. Mixing concrete with a machine is more cost-effective, produces concrete of higher quality in a shorter amount of time, and does it at a faster rate. Concrete is typically mixed with one of two different types of mixers:

(1) Continuous mixers: These kind of mixers are utilised in major construction projects that require a large quantity of concrete to be mixed continuously [35]. The feeding of the mixing plant is performed in a manner that is largely automatic. The device calls for vigilant attention and monitoring at all times [40].

(2) Batch mixers: In this sort of mixer, the required amounts of materials are put into the hopper of a drum [18]. The components are then combined by a series of blades that are contained within the drum. These are an extension of the near drum type and the tilting drum type [33].

8

(i)Pan mixers: These are appropriate for extremely dry concrete mix, and they are utilised in manufacturing facilities for the production of precast products.

(ii) Paver mixers: These are employed for laying concrete for pavements, and their name comes from their function [7].

(iii) Double drum mixers: These have the ability to finish mixing at one end of the drum and discharge the product while the other end of the drum is used to load new ingredients and begin the mixing process for the next batch [41].

5.   Transportation of concrete

Concrete should be transported from the place where it is mixed to the place where it is deposited in the shortest time possible. This is done to ensure that there is no segregation or loss of ingredients and that the material maintains its workability while it is being placed in position [19]. The following are some of the several methods that can be utilised: steel pans, wheel barrows, dumpers, trippers, truck mixers, hoists, cableways, cranes, and pumping. Because of this, the transportation of fresh concrete and concrete in general is an essential operation [37]. Concrete is not a homogenous combination; rather, it is a mixer of components that differ from one another in size and specific gravity. This results in the phenomenon of segregation [24]. Therefore, as soon as the concrete is discharged from the mixer, both internal and external forces begin to act to separate the various ingredients into their respective categories [31]. When overly wet concrete is placed in forms that are too restrictive, the coarser and heavier particles have a tendency to sink, while the finer and lighter components have a tendency to rise [42]. Therefore, the equipment for carrying concrete from the mixer to the form should be designed in such a way that it produces concrete that is consistent throughout and uniform at the form [22]. The method of carrying should not cause segregation, excessive drying or stiffening, nor should

9

it require wetter concrete than is necessary at the form for placement. Neither should it produce excessive drying [43]. When transferring concrete from one conveyance to another, in order to prevent the concrete from becoming separated into its constituent parts, the usage of hoppers, baffles, and brief vertical drops should be implemented through a pipe to the centre of the receiving container [8]. This misconception, that segregation that occurs during transportation can be fixed by compacting the material, is not accurate [31]. The following is a list of the various modes of transportation:

1. Pan method:

A). The use of a wheelbarrow as a mode of transportation.

B). Power barrows.

1. The use of dump trucks for transportation.

2. Transportation using tractors and trailers

3. Transportation utilising jubilee trucks and monorail systems.

4. Transportation by means of a hoist

5. Transportation using a crane

6. The movement of concrete by an elevating tower and a series of chutes

7. Moving concrete using cable cars suspended from the ceiling

Ready mix concrete is a type of concrete that may be placed directly from a central factory where it was mixed [14]. It offers a number of benefits, including the fact that certain types of concrete are suitable for crowded building sites, such as those used for road construction in areas where there is limited room for mixing and stocking aggregates [29]. It is possible to exercise greater control over the mixing process when using ready-mix concrete. The most significant detriment is the expensive price [44].

10

It has been discovered that transporting concrete using pumps is helpful in the following scenarios:

• At congested sites where the mixing plant cannot be moved close to the point where concrete is being placed [39].

• At locations where there is a lack of available space for storing aggregate, which makes it impossible to do so.

Concrete can be transported via this approach to locations across a large area that are, under normal circumstances, difficult to reach [42].

There is no segregation in concrete that has been pumped. Pumping the concrete eliminates the need for double handling because it sends the concrete straight from the mixer to the location [20]. The limitation of the equipment used for carrying and putting does not slow down the placing process, which means it can proceed at the same rate as the output of the mixer [45].

When deciding how to transport concrete, both the scale of the project and the quantity of the material that must be moved must be taken into consideration [1]. When working on big areas or at heights that require concreting, the transportation method of pumping is brought back into use. The diameter of the steel tubes that make up the pipeline ranges from 100 to 125 millimetres, and their length is 3 metres [34]. It is a strategy that is both incredibly quick and quite effective [18].

6.   Arrangement of the concrete

After the concrete has been mixed, it should be put and compacted as soon as possible. The most important things to do in the field are placing things and compacting them [39]. Even if the concrete is mixed perfectly, if these processes are not carried out with extreme caution, the end result could be a very poor job. If it does happen, the most important thing is to avoid division [27]. When concrete is being poured into deep forms, it is typically permitted to reach the full depth of the form regardless of its height. As a consequence, this leads to segregation as well as damage to the form and embedded fixings. Reinforcement and forms that are placed above the level of placement receive a coating of mortar, which may have time to dry before the concrete reaches that level [32]. In the event that horizontal layers are to be created, the concrete should be deposited as close as possible to where it will ultimately be used. In the event that concrete needs to be placed in a slab, it needs to be dropped onto the face of concrete that was already there. Permeating consolidation of the layer using vibration or any other method that is suited in situations where concrete is not to be laid on a slope that could potentially slip [41]. When placement is begun at the steepest part of the slope, vibration has a tendency to push the material lower down the slope [6].

It is essential to ensure that the arrangement for placement is carried out in such a manner that the mixed mass is used within thirty minutes. This is required in order to avoid initial set. The dust should be wiped off of the concrete shuttering before it is laid down [44]. The concrete needs to be crushed appropriately after being layered and placed down [40].

Caution should be exercised as follows while placing concrete:

1) When using reinforced concrete, the thickness of the concrete can range anywhere from 15 to 30 centimetres, and it needs to be layered in a horizontal and consistent fashion.

2) In order to avoid the concrete from segregating, it should never be thrown from a height of more than one metre.

3). Walking on freshly constructed concrete should never be permitted under any circumstances.

4. The placement of concrete should be halted whenever there is rain in the forecast.

5. The laitance that was placed on top of the old concrete needs to be removed before the new concrete can be placed.

6) Concrete needs to be laid in one continuous layer in order to prevent the formation of an uneven surface.

7). It is important to avoid disturbing the alignment of the reinforcement and the concrete.

8). When working with RCC slabs, the process of concreting should begin widthwise from the end.

9). The removal of grease, oil, and dry cement is recommended in order to generate a better bond between the concrete and the reinforcement.

7.   Segregation

In order for a concrete mix to maintain its consistency during the placement and transportation processes, it must avoid separating into individual components and bleeding [19]. The process of separating out the components of a concrete mix, to the point where the mixture is no longer in a homogenous state, is referred to as segregation. Only the consistent, homogenous mix is capable of being thoroughly compacted [5].

The sorting process is dependent on the processes of handling and placement. Because of its propensity to separate, the quantity of coarse aggregate used, and the increased slump, Reducing the height of the drop created by the concrete is one way to help limit the likelihood for groups to form [42].

b. Not utilising the vibration as a technique of distributing a huge mass of concrete across a wide area in order to create a flat surface.

c. Lessening the vibration that has been going on for an extended period of time, since the coarse aggregate has a tendency to sink to the bottom and the scum would rise to the surface.

d. The addition of a little amount of water, which boosts the cohesiveness of the mixture [34].

Because the solid particles in the mixture are unable to absorb all of the mixing water while the mixture is being compacted, bleeding occurs as a result of the water in the mixture rising to the surface [7]. This is because the solid particles are unable to prevent the water from evaporating. Because of the bleeding, the top layer of the put concrete develops a porous, fragile, and non-durable concrete layer [43]. In the event of lean mixes, bleeding may result in the formation of capillary channels, which will increase the permeability of the concrete. When the concrete is laid down in several layers, and each layer is compacted after a specific amount of time has passed before the next layer is laid, bleeding may cause a plane of weakness to form between two of the layers [44]. Before adding a new layer, the previous one should be brushed and washed in order to remove any laitance that may have occurred. It is important to prevent compacting the surface too much [45].

8.   Conclusion

The four major components of concrete are cement, sand, stone aggregates, and water. Reinforced cement concrete, or RCC for short, is created by combining a cage formed of steel rods with the concrete mix. As a result, the substance is created. When subjected to stresses during limited-type handling operations, fresh concrete should be stable and not separate or bleed during transportation and placing. Variations in the quality of the concrete have been discovered to be partially linked to flaws that happened during the batching process. The mixing method should be carefully chosen, and the following factors should be considered. Maintaining the mixer in such a way that it produces enough concrete necessitates doing it in an ordered and regular manner. Concrete should be carried as quickly as possible from the point where it is mixed to the point where it is deposited. This is done to guarantee that no ingredients are separated or lost, and that the material retains its workability while being positioned in position.

References

[1]   R. Garg, R. Garg, and N. O. Eddy, "Microbial induced calcite precipitation for self-healing of concrete: a review," *J. Sustain. Cem. Mater.*, 2022, doi: 10.1080/21650373.2022.2054477.

[2]   H. Liu, Q. Li, D. Su, G. Yue, and L. Wang, "Study on the influence of nanosilica sol on the hydration process of different kinds of cement and mortar properties," *Materials (Basel).*, vol. 14, no. 13, 2021, doi: 10.3390/ma14133653.

[3]   S. Mandal, J. K. Singh, D. E. Lee, and T. Park, "Effect of phosphate-based inhibitor on corrosion kinetics and mechanism for formation of passive film onto the steel rebar in chloride-containing pore solution," *Materials (Basel).*, vol. 13, no. 16, pp. 7–9, 2020, doi: 10.3390/MA13163642.

[4]   E. Najaf, M. Orouji, and S. M. Zahrai, "Improving nonlinear behavior and tensile and compressive strengths of sustainable lightweight concrete using waste glass powder, nanosilica, and recycled polypropylene fiber," *Nonlinear Eng.*, vol. 11, no. 1, pp. 58–70, 2022, doi: 10.1515/nleng-2022-0008.

[5]   Antoni, J. G. Halim, O. C. Kusuma, and D. Hardjito, "Optimizing Polycarboxylate Based Superplasticizer Dosage with Different Cement Type," *Procedia Eng.*, vol. 171, pp. 752–759, 2017, doi: 10.1016/j.proeng.2017.01.442.

[6]   H. Sharma, R. Garg, D. Sharma, M. umar Beg, and R. Sharma, "Investigation on Mechanical Properties of Concrete Using Microsilica and Optimised dose of Nanosilica as a Partial Replacement of Cement," *Int. J. Recent Reasearch Asp.*, vol. 3, no. 4, pp. 23–29, 2016.

[7]   R. Garg, R. Garg, B. Chaudhary, and S. Mohd. Arif, "Strength and microstructural

15

analysis of nano-silica based cement composites in presence of silica fume," *Mater. Today Proc.*, vol. 46, pp. 6753–6756, 2020, doi: 10.1016/j.matpr.2021.04.291.

[8]   H. B. Tran, V. B. Le, and V. T. A. Phan, "Mechanical properties of high strength concrete containing nano sio2 made from rice husk ash in Southern Vietnam," *Crystals*, vol. 11, no. 8, 2021, doi: 10.3390/cryst11080932.

[9]   Rishav Garg, Manjeet Bansal, and Yogesh Aggarwal, "Split Tensile Strength of Cement Mortar Incorporating Micro and Nano Silica at Early Ages," *Int. J. Eng. Res.*, vol. V5, no. 04, pp. 16–19, 2016, doi: 10.17577/ijertv5is040078.

[10]   Atta-ur-Rehman, A. Qudoos, S. H. Jakhrani, H. G. Kim, and J.-S. Ryou, "Influence of Nano-silica on the Leaching Attack upon Photocatalytic Cement Mortars," *Int. J. Concr. Struct. Mater.*, vol. 13, no. 1, p. 35, Dec. 2019, doi: 10.1186/s40069-019-0348-x.

[11]   D. Prasad Bhatta, S. Singla, and R. Garg, "Experimental investigation on the effect of Nano-silica on the silica fume-based cement composites," *Mater. Today Proc.*, vol. 57, pp. 2338–2343, 2022, doi: 10.1016/j.matpr.2022.01.190.

[12]   P. Smarzewski, "Influence of silica fume on mechanical and fracture properties of high performance concrete," *Procedia Struct. Integr.*, vol. 17, pp. 5–12, 2019, doi: 10.1016/j.prostr.2019.08.002.

[13]   A. Singh, S. Singla, R. Garg, and R. Garg, "Performance analysis of Papercrete in presence of Rice husk ash and Fly ash," *IOP Conf. Ser. Mater. Sci. Eng.*, vol. 961, p. 012010, Nov. 2020, doi: 10.1088/1757-899X/961/1/012010.

[14]   H. T. Le and H. M. Ludwig, "Effect of rice husk ash and other mineral admixtures on properties of self-compacting high performance concrete," *Mater. Des.*, vol. 89, pp. 156–166, 2016, doi: 10.1016/j.matdes.2015.09.120.

[15]   J. J. Chen, P. L. Ng, L. G. Li, and A. K. H. Kwan, "Production of High-performance Concrete by Addition of Fly Ash Microsphere and Condensed Silica Fume," *Procedia Eng.*, vol. 172, pp. 165–171, 2017, doi: 10.1016/j.proeng.2017.02.045.

[16]   M. Benaicha, A. Hafidi Alaoui, O. Jalbaud, and Y. Burtschell, "Dosage effect of superplasticizer on self-compacting concrete: Correlation between rheology and strength," *J. Mater. Res. Technol.*, vol. 8, no. 2, pp. 2063–2069, 2019, doi: 10.1016/j.jmrt.2019.01.015.

[17]   G. M. Fani, S. Singla, R. Garg, and R. Garg, "Investigation on Mechanical Strength of Cellular Concrete in Presence of Silica Fume," *IOP Conf. Ser. Mater. Sci. Eng.*, vol. 961, no. 1, p. 012008, Nov. 2020, doi: 10.1088/1757-899X/961/1/012008.

[18]   C. Xupeng, S. Zhuowen, and P. Jianyong, "Study on Metakaolin Impact on Concrete Performance of Resisting Complex Ions Corrosion," *Front. Mater.*, vol. 8, no. December, pp. 1–12, 2021, doi: 10.3389/fmats.2021.788079.

[19]   R. Garg, T. Biswas, M. D. Alam, A. Kumar, A. Siddharth, and D. R. Singh, "Stabilization of expansive soil by using industrial waste," *J. Phys. Conf. Ser.*, vol. 2070, no. 1, p. 012238, Nov. 2021, doi: 10.1088/1742-6596/2070/1/012238.

[20]   Y. C. Ersan, "Overlooked Strategies in Exploitation of Microorganisms in the Field of Building Materials," Springer Singapore, 2019, pp. 19–45. doi: 10.1007/978-981-13-0149-0_2.

[21]   L. Senff, D. Hotza, and J. a Labrincha, "Effect of diatomite addition on fresh and hardened properties of mortars investigated through mixture experiments," *Adv. Appl. Ceram.*, vol. 110, no. 3, pp. 142–150, Apr. 2011, doi: 10.1179/1743676110Y.0000000009.

[22]   R. Garg, R. Garg, and N. O. Eddy, "Influence of pozzolans on properties of cementitious materials: A review," *Adv. Nano Res.*, vol. 11, no. 4, pp. 423–436, 2021, doi: 10.12989/anr.2021.11.4.423.

[23]   A. Nazari, S. Riahi, S. Riahi, S. F. Shamekhi, and A. Khademno, "Mechanical properties of cement mortar with Al2O3 nanoparticles," *J. Am. Sci.*, vol. 6, no. 4, pp. 94–97, 2010.

[24]   K. Kumar, M. Bansal, R. Garg, and R. Garg, "Mechanical strength analysis of fly-ash based concrete in presence of red mud," *Mater. Today Proc.*, vol. 52, no. xxxx, pp. 472–476, 2022, doi: 10.1016/j.matpr.2021.09.233.

[25]   U. Ubinayaa, D. Chetha, S. Chathuska, N. Praneeth, R. Vimantha, and K. K. Wijesundara, "Improving the properties of concrete using carbon nanotubes," *SAITM Res. Symp. Eng. Adv.*, vol. 2014, p. 4, 2014.

[26]   T. Parhizkar, A. M. R. Ghasemi, and A. A. Ramezanianpour, "Properties of the New Type of HPC in Simulated Conditions of Persian Gulf".

[27]   R. Garg *et al.*, "Mechanical strength and durability analysis of mortars prepared with fly ash and nano-metakaolin," *Case Stud. Constr. Mater.*, vol. 18, no. December 2022, p. e01796, 2023, doi: 10.1016/j.cscm.2022.e01796.

[28]   R. P. S. Kushwah and O. Prakash, "Utilization of 'marble slurry' in cement mortar," *Int. J. Eng. Sci. Res. Technol.*, vol. 5, no. 9, pp. 640–645, 2016, doi: 10.5281/zenodo.155087.

[29]   J. Yu, M. Zhang, G. Li, J. Meng, and C. K. Y. Leung, "Using nano-silica to improve mechanical and fracture properties of fiber-reinforced high-volume fly ash cement mortar," *Constr. Build. Mater.*, vol. 239, p. 117853, Apr. 2020, doi: 10.1016/j.conbuildmat.2019.117853.

[30]   A. Singh, A. Garg, and R. Garg, "Influence of Nano-Silica and Ground Granulated Blast Furnace Slag on Cement Using Statistical," 2018.

[31]   D. G. Leo Samuel, K. Dharmasastha, S. M. Shiva Nagendra, and M. P. Maiya, "Thermal comfort in traditional buildings composed of local and modern construction materials," *Int. J. Sustain. Built Environ.*, vol. 6, no. 2, pp. 463–475, 2017, doi: 10.1016/j.ijsbe.2017.08.001.

18

[32]	B. B. Das and A. Mitra, "Nanomaterials for Construction Engineering-A Review," *Int. J. Mater. Mech. Manuf.*, vol. 2, no. 1, pp. 41–46, 2014, doi: 10.7763/ijmmm.2014.v2.96.

[33]	Y. Qing, Z. Zenan, K. Deyu, and C. Rongshen, "Influence of nano-SiO2 addition on properties of hardened cement paste as compared with silica fume," *Constr. Build. Mater.*, vol. 21, no. 3, pp. 539–545, Mar. 2007, doi: 10.1016/j.conbuildmat.2005.09.001.

[34]	M. Seifan, A. Ebrahiminezhad, Y. Ghasemi, A. K. Samani, and A. Berenjian, "The role of magnetic iron oxide nanoparticles in the bacterially induced calcium carbonate precipitation," *Appl. Microbiol. Biotechnol.*, vol. 102, no. 8, pp. 3595–3606, 2018, doi: 10.1007/s00253-018-8860-5.

[35]	Y. Kocak, "A study on the effect of fly ash and silica fume subsituted cement paste and mortars," *Sci. Res. Essays*, vol. 5, no. 9, pp. 990–998, 2010.

[36]	R. Sharma, D. Sharma, and R. Garg, "Development of Self Compacted High Strength Concrete using Steel Slag as Partial Replacement of Natural Fine Aggregate," pp. 96–99, 2016.

[37]	N. Stevulova, I. Schwarzova, V. Hospodarova, and J. Junak, "Implementation of waste cellulosic fibres into building materials," *Chem. Eng. Trans.*, vol. 50, pp. 367–372, 2016, doi: 10.3303/CET1650062.

[38]	A. Albidah, M. Alghannam, H. Abbas, T. Almusallam, and Y. Al-Salloum, "Characteristics of metakaolin-based geopolymer concrete for different mix design parameters," *J. Mater. Res. Technol.*, vol. 10, pp. 84–98, 2021, doi: 10.1016/j.jmrt.2020.11.104.

[39]	A. Garg, A. Singh, and R. Garg, "Effect of Rice Husk Ash & Cement on CBR values of Clayey soil," *Int. J. Eng. Res. Appl.*, vol. 3, no. 1, pp. 1710–1717, 2013.

[40]	K. Kumar, M. Bansal, R. Garg, and R. Garg, "Penetration and strength analysis of pervious concrete," *J. Phys. Conf. Ser.*, vol. 2070, no. 1, 2021, doi: 10.1088/1742-

6596/2070/1/012244.

[41]   M. Jalal, A. R. Pouladkhan, H. Norouzi, and G. Choubdar, "Chloride penetration, water absorption and electrical resistivity of high performance concrete containing nano silica and silica fume," *J. Am. Sci.*, vol. 8, no. 4, pp. 278–284, 2012.

[42]   J. Liu, H. Jin, C. Gu, and Y. Yang, "Effects of zinc oxide nanoparticles on early-age hydration and the mechanical properties of cement paste," *Constr. Build. Mater.*, vol. 217, pp. 352–362, 2019, doi: 10.1016/j.conbuildmat.2019.05.027.

[43]   R. Singh and S. Goel, "Experimental investigation on mechanical properties of binary and ternary blended pervious concrete," *Front. Struct. Civ. Eng.*, vol. 14, no. 1, pp. 229–240, 2020, doi: 10.1007/s11709-019-0597-4.

[44]   E. Molaei Raisi, J. Vaseghi Amiri, and M. R. Davoodi, "Mechanical performance of self-compacting concrete incorporating rice husk ash," *Constr. Build. Mater.*, vol. 177, pp. 148–157, 2018, doi: 10.1016/j.conbuildmat.2018.05.053.

[45]   H. S. Gökçe, D. Hatungimana, and K. Ramyar, "Effect of fly ash and silica fume on hardened properties of foam concrete," *Constr. Build. Mater.*, vol. 194, pp. 1–11, 2019, doi: 10.1016/j.conbuildmat.2018.11.036.